The Paperclip Mindset

Frank Karsten

The Paperclip Mindset

Radical simplicity for project success

The Paperclip Mindset
Radical simplicity for project success

First edition: May 30th 2022
Current edition: 2.3 (April 19th 2026)
Cover image: Shutterstock
Website: paperclip.one
frank@frankkarsten.com

Copyright of images
Cover image: purchased at Shutterstock
Author image: copyright of the author
All other images: Creative Commons or unknown

ISBN-13: 9798439560509 - Paperback

About the author

Frank Karsten is an experienced IT professional with a rich background in both startup ventures and traditional IT projects. He has advised numerous companies on decluttering their stalled projects.

He is author of two other books that deal with societal and political issues, Beyond Democracy (2012) and The Discrimination Myth (2019).

Acknowledgements

I would like to thank Remy Tensen for helping me to write this book. He and I have collaborated on many IT projects and have discussed their principles and dynamics regularly and with great passion. From the moment we started working together, simplicity was always our main objective. The ideas presented in this book are in large part thanks to our discussions.

I would also like to thank Joost van Hoof for his ideas on improving IT projects. Joost and I collaborated on a startup project in 2014. Thanks to our complementary skills we were able to successfully deliver the project within a mere six weeks. It also helped greatly that we worked in the same room two meters apart. From that project on I started documenting my thoughts on better IT projects, which eventually resulted in this book.

Table of contents

About the author	**5**
Acknowledgements	**6**
Table of contents	**7**
Introduction	**11**
What is Paperclip?	**17**
The benefits	**21**
The Complexity Monster	**23**
When is a project failed?	**29**
What goes wrong?	**31**
Specific IT challenges	**35**
The 4 project constraints	**39**
The 4 Paperclip Objectives	**41**
1 - Minimize time	43
2 - Minimize target	47
3 - Minimize team	51
4 - Minimize technology	57
Why time is the best constraint	**61**
'Good Enough' is Perfect (for now)	**63**
The benefits of working in-person	**65**
A simple simplification technique	**69**
The importance of pushback	**71**
Cases	**75**
Case 1: Live now, pay later	75
Case 2: Stop automating everything!	76
Case 3: Fake it till you make it	77
Case 4: Who needs Excel?	78
Case 5: Forget the MailChimp API	79

Case 6: Beware of 'Layer Love'	80
Case 7: Who needs an API?	84
Case 8: Cutting out the nonsense	85
Case 9: Avoiding the time zone hell	87
Case 10: Let's create luxury problems!	89
Case 11: Avoiding the flight forward	91
FAQ	**95**

"It always takes longer than you expect, even when you take into account Hofstadter's Law"

(Douglas) Hofstadter's Law (1945)

Introduction

IT projects are notorious for their high fail rate. In the newspaper we can often read about IT projects that were scrapped because of huge time and money overruns. One of the biggest failures was the Obamacare website on which Americans could sign up for it. It cost no less than $840 million and was extremely sluggish. But luckily they got it working still.

The Dutch RaboBank was not so fortunate. It spent 200 million on its company wide LAURA system before management considered it beyond repair and subsequently ditched it. But no matter how big or small the IT project, they tend to fail in similar fashion. Overcomplexity is often the main cause.

The problem with IT projects is that this complexity is even worse than in other sectors. Unlike shoes or bridges, software is invisible and abstract. As an outside observer you can hardly assess its complexity. Even those involved in the project are often unaware of it.

John Wanamaker, an American retailer and pioneer in marketing once said *"Half the money I spend on advertising is wasted, and the trouble is I don't know which half."* Most of us probably experienced something similar when realizing a project. We waste enormous amounts of time, money and effort on the wrong things. The problem is that we tend to only know in hindsight what those things were. Wouldn't it be great if we could identify *in advance* what elements should *not* be done?

This book deals with helping you to do that, and the gains can be huge. Implementing a functional element two, four or sometimes even thirty times faster can result in huge savings. Spending just one hour of thinking that deletes a three-week task delivers a 120× Return On Investment.

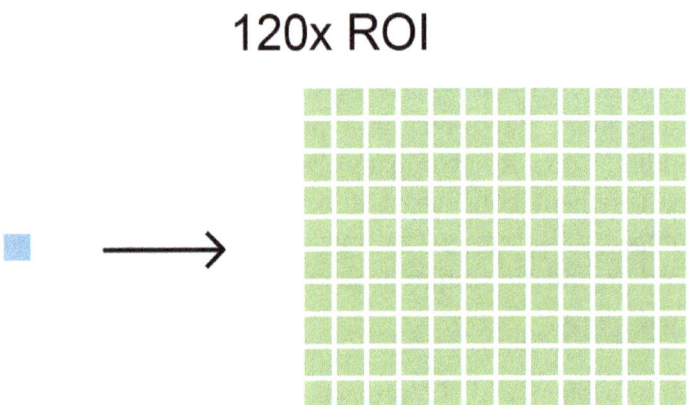

Do less, achieve more
This book also helps you and others to keep you on edge, to keep each other sharp in saving time and money. It will help you to enjoy your work more, for you will better understand what the project dynamics are, and because you will achieve more while doing less.

But no matter how much we are aware of the pitfalls of the projects we do, and the difficulties to assess what nót to do, we always tend to do too much. The American philosopher Douglas Hofstadter said it well in a self-referencing law that is named after him: *"It always takes longer than you expect, even when you take into account Hofstadter's Law"*.

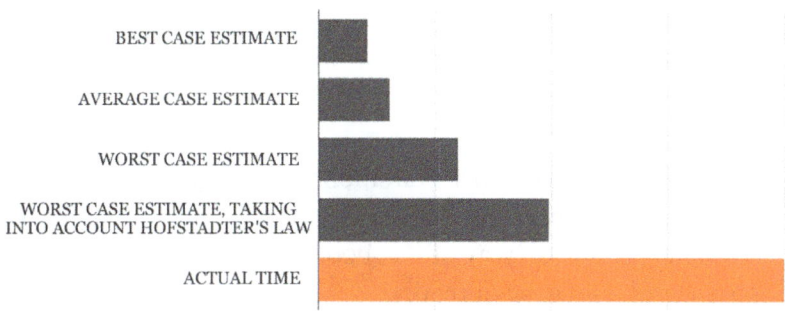

But realizing Hofstadter's rather pessimistic outlook should not discourage us, because it means we are less taken by surprise. It makes us better understand what we are up against, which can make us more energetic and optimistic about our chances to succeed.

The second law of thermodynamics
So even if you fail to plan correctly, don't just blame yourself, blame the universe too. The universe is forcing the second law of thermodynamics on all of us, namely the tendency towards entropy, chaos. Given all the chaos, it's a miracle that we as a humanity ever came to this advanced state.

And let's be real, most projects are chaotic, and tend towards more complexity over time, not less. The trick is to make them less chaotic, more structured and more purposeful. Once you better master it, you come to enjoy taming projects, no matter how small.

Not a trick but an art
After much deliberation, and a lot of both failed and successful projects I identified four main objectives (the 4 Ts, which I cover in chapter *The 4 Paperclip Objectives*) that will help you harness your IT project. Please note that this is not a simple trick with easy answers. It is an art that requires continuous practice and reflection.

I agree with Einstein's statement that if you can't explain something simply, you probably don't understand it well enough. That is why this book is deliberately concise, structured, and easy to read.

I also included various examples from my personal experience. Some of them saved a few weeks of work, others possibly even 9 to 12 months. Hopefully you will experience similar savings in time, effort and money.

"If you can't explain it simply, you don't understand it well enough"

Albert Einstein (1879 – 1955)

What is Paperclip?

Paperclip is the art of doing less and achieving more. It is not a method but a mindset to continuously identify what NOT to do. It's about focussing on the right things, and doing them in the simplest way. This book hopes to convey three things to you:

1. Understanding the threat of the Complexity Monster
2. Creating a framework to incentivize simplicity
3. Identifying what NOT to do by providing examples

Paperclip is about the brutal pursuit of simplicity, that it should be the main design objective for all aspects of the project, in planning, team size, targets, communication and technology. Even though most people would agree with this goal, few know what simplicity would look like and how to achieve it. How can we know we have reached maximum simplicity?

Paperclip works independently of what kind of project management tool you use, or which programming language you employ. Many of the ideas *even* work for projects outside of IT.

Paperclip can be considered *Lean* (in the senses of minimalistic) but is quite different from modern software development methodologies like Lean Software Development, Scrum or Agile, although it shares certain elements. While Lean, Scrum and Agile focus on how to do things better, Paperclip focuses on how to understand what not to do at all. I have been part of Agile and Scrum projects that went way overboard and became overengineered, because both methodologies don't help much in fostering and understanding simplicity.

Simplifying your IT project comes with all kinds of indirect benefits. Tony Hoare, winner of the the prestigious Turing award in computer science, once stated: "The price of reliability is the pursuit of the utmost simplicity". But beside reliability there are other advantages too:

- Debuggable code
- Low cost / low maintenance
- High operating speed (fast code)
- Transferability of code

When something is very complicated we tend to call it 'rocket science'. *Paperclip* is about 'reverse rocket science'. The trouble with a rocket, besides its intrinsic complexity, is that it becomes exponentially complex by adding more payload. For every one kilogram of extra payload you need to add at least nine kilograms of extra fuel, again leading to even more requirements. It easily causes what I call a 'Chain Reaction of Negatives', a downward spiral.

Luckily, this effect of exponential complexity works in reverse too, for if we simplify things it often leads to a chain reaction of benefits, an upward spiral. When you remove elements, the need for certain other elements often falls away. Reducing the goal can result in a smaller team working on it, which results in less communication, less time and lower costs.

Since *Paperclip* deals with IT projects you might think it's also about programming or database design. But it's not. And although simplicity works for all tasks, processes and projects, in all industries and at all times, this book focuses solely on web projects. It's not about the serial production for three million shoes, like in a factory. And it's not about the commercialization of IT projects (successfully delivered web projects can still fail because of bad marketing). Neither is it about IT projects that have failed for financial or organizational reasons.

"The essence of strategy is choosing what not to do."

Michael Porter (1947)
American business guru

The benefits

In the past, projects were fuzzy processes for me which I did not understand well. But now I consider myself pretty competent, both consciously and unconsciously. I have experienced great benefits by applying the Paperclip Mindset:

1. Lower project costs
2. Faster delivery
3. Better results

Enjoyable work
An additional benefit is that I enjoy work more now, for I am more conscious of what I am doing and therefore more successful. Customers like it too, for the Paperclip Mindset is a set of ideas that they can understand easily without much technical jargon.

But before I overpromise, remember that it's an art, not a trick. There will never be an easy general answer to the project difficulties you experience. You will never stop making these mistakes, but you will always learn to make fewer. In order to get maximum benefit from the Paperclip Mindset you will need to cultivate the Paperclip way of thinking among your colleagues and customers. I have noticed that customers are very receptive to these ideas, and I am confident you will have similar experiences.

"It's not about being right, but about being less wrong."

Elon Musk

The Complexity Monster

Who is this monster, which I call the Complexity Monster, that makes our projects and tasks so overly complex?

First, it is our natural inclination to make things more complex than necessary, to continuously add things. Our brain is wired to complicate stuff. We generally do this in the name of positive sounding words, like completeness, quality, security, ambition, eventualities, beauty, etc. Since these terms sound appealing it makes it very hard to argue against the added complexity. For you could be perceived as lazy, unambitious or careless for challenging them, whereas in reality you might just want to be more productive and meet targets.

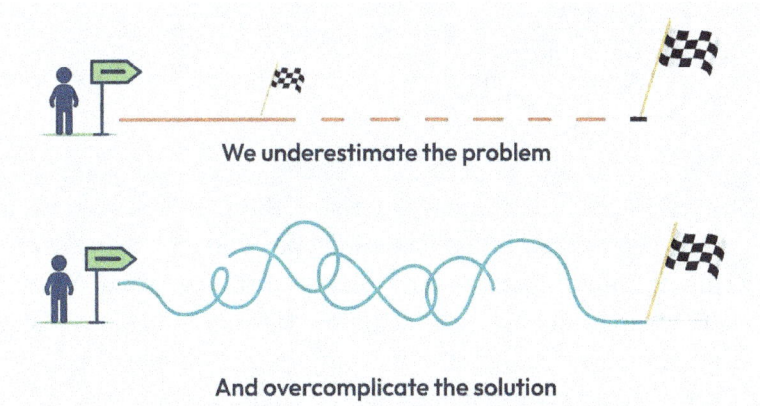

We underestimate the problem

And overcomplicate the solution

The second part of Complexity Monster is what I call the Complexity Chain Reaction. This means that by adding one aspect it often results in other new elements being required too, adding even more to the complexity. For instance, adding just a single team member might trigger the need for more office space, extra meetings, a new HR manager, etc.

To complicate or to simplify
In general, 'to complicate' means to add stuff, and 'to simplify' means to remove stuff. We tend to think that 'stuff' means 'functional elements', like a button or a screen, but they can be so many other things too, like:

- Team members
- Working locations
- Communication tools
- Options to choose from
- Gadgets
- Planning tools
- Beauty
- Technology
- Decision-makers
- Quality

By adding these things you risk delays, misunderstandings, meetings, uncertainties, dependencies and *Analysis Paralysis*. This of course does *not* mean that nothing can be added, but that we must be aware that these additions come at a cost, and that the gains are great if we can keep it simple. So don't ask twenty people what the best color is, stop aiming for perfection before releasing, quit adding niceties and stop writing elaborate emails.

But even if we better understand the Complexity Monster we must remember that the creature never sleeps and can not be killed. Luckily though, it can be harnessed.

Unfortunately, minimalism is not very fashionable or sexy. If one proposes to do less, it often appears to others that one is lazy or unambitious. Once I audited a bogged down IT project and I asked the team why they had added a layer with XML (a standard data format), which I considered unnecessary. One software engineer scoffed at me and said: "But XML is not difficult, is it?" As if he wanted to suggest I must be unintelligent for raising the question. As if I didn't understand XML and would therefore rather get rid of it. I was shocked by his response and I have always remembered it. For the main question is never whether it is difficult, practical, best practice, or the industry standard, the main question is whether it is really needed.

The graph below shows how complexity grows exponentially with functionality (or number of elements). The green line is the minimum amount of complexity for that particular functionality to work. All the complexity underneath it is called *essential*, while any level above the green line is *accidental*, something you don't want but generally happens anyway.

Given a certain amount of functionality, you would ideally end up on the green line, but in practice anywhere in the lower left corner is already very good. The graph also indicates that slashing functionality is often just as necessary as slashing complexity.

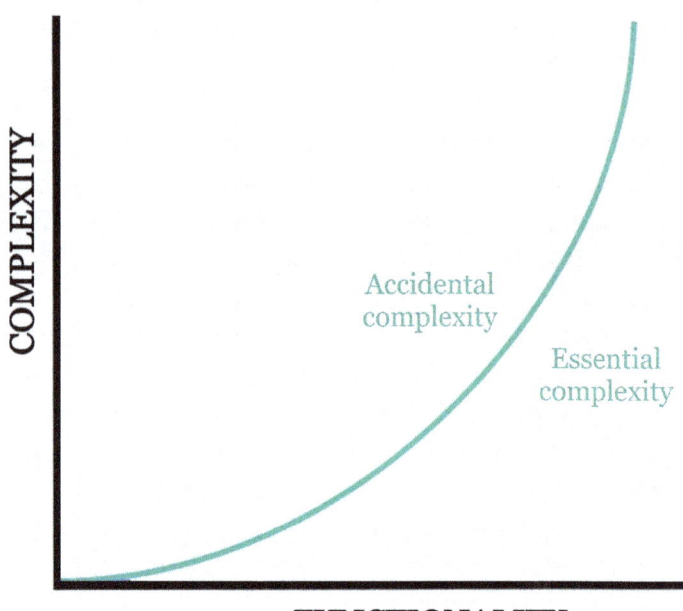

Why engineers like complexity

Many technical professionals are prone to complicate things, even more than the average person. I think there are three reasons for that.

1. If you make it complex you make yourself indispensable. You can't be fired.
2. You can indirectly boast about your technical skills and intelligence.
3. Adding a new technology to your skill stack will look good on your resume.

So adding and mastering complexity gives you status and a higher, more stable income, even though the chances of successfully completing the project are smaller. This is a perverse incentive that must be countered.

Make simplicity sexy again

Since our brain is wired for complexity, we must be trained to simplify. The trick is to make simplicity sexy again within the team. To incentivize all team members to come up with simpler solutions, and to praise everyone who was able to think of a short-cut or a handy solution. This requires continuous evangelizing. Elon Musk once remarked that he fires employees who do not innovate enough. I suspect Musk considers simplification to be innovation too.

"Simplicity is the ultimate sophistication"

Leonardo da Vinci (1452 - 1519)

When is a project failed?

In 2009 the computer magazine ZdNet reported that, according to research, more than half of all IT projects fail. But what does it mean to fail? The answer is not so clear, but that's ok. For me a failed project is a project that is way over time, over budget or in which the deliveries turned out very unusable or very faulty. I would not consider a project 'failed' that is just 10% over budget or over schedule. In fact, it probably is a huge success.

However, even a failed project is not necessarily a disaster. It could still work and could still not break the bank. Nearly all (IT) projects go over time or over budget. Even failed projects that took double the projected time and budget can turn out successfully, simply because the resulting IT system is functional and serves its users.

"Intelligence is the ability to avoid doing work, yet getting the work done."

Linus Torvalds (1969)
Creator of the Linux OS

What goes wrong?

Pretty much anything can and does go wrong during an IT project: expectations, communication, planning, motivation, technology, funding, etc. When doing my first projects I wondered why the universe had uniquely selected me to be so unfortunate to become involved with such cumbersome projects. Only later on I noticed that the problems are universal, and that other projects fared even worse!

Below you see a comical depiction of the failures in a regular IT project (visit www.paperclip.one for a more readable image).

The various individuals described in the cartoon experience great difficulty in understanding each other. Mind you, even the customer does not seem to really know what kind of swing he wants, and that is often true to reality.

Communication between people is often flawed and ambiguous. And even if that goes well, the planning can and generally does go wrong. But even if communication and planning are successful, prioritizing features and tasks tends to fail. And if all technical stuff is right in place the team members might be lacking motivation. In short, there are so many things that can go wrong, and it's important to be aware of it.

What I learned from this is to be as cooperative and communicative as possible as a team member. I also realized I need to get rid of my ego during projects, to be goal oriented and to leave my pet subjects behind. I came to understand that reaching the goal is difficult enough, even when all requirements are in place and everyone works hard and is intelligent and capable. Acknowledging this fact made me more humble.

These challenges apply to all projects, but let's see what the specific IT challenges are.

"The cheapest, fastest, and most reliable components are those that aren't there"

Gordon Bell (1934)
Computer scientist

Specific IT challenges

Software development is even more prone to complexity than other technologies because of certain inherent characteristics.

1) Abstraction

Buildings are physical things, but software is abstract and is not physical in essence. It is therefore difficult to comprehend for the average person. If buildings were designed and built the way software is designed a six year old child would notice the absurdity of many designs. It would appear like the stairwell in this famous Escher drawing.

A building designed by a software architect?

But because of the intrinsic abstractness of software even experts find it difficult to notice whether a system is overengineered and overly complex. And if we can not assess complexity it's more difficult to contain.

2) Moore's Law

In 1965 Gordon Moore, co-founder of chipmaker Intel, coined his famous Moore's Law. It is not a real law but just his correct observation that the number of transistors on an integrated circuit (like a computer processor) roughly doubles about every two years. This means that software runs faster and faster over time. This seems like a big plus but for software development it's also a huge danger. For it causes bloated over-engineered software designs to still run almost as quickly as well-designed software systems. Faster chips obscure bad software. In other words, Moore's Law is forgiving towards complexity and bad design.

Computer scientist Niklaus Wirth noticed this adverse trend too and coined Wirth's Law in 1995: "Software is getting slower more rapidly than hardware is becoming faster".

3) Customer lock-in

Even when the customer owns the copyright to the software it is still extremely costly to change supplier. Software is often complex and not very well documented. And even if the system is well documented it can not easily be transferred to a new development team at another company.

This customer lock-in creates a Premium on Failure for the IT supplier. Even when it does not deliberately want to abuse its rather exclusive position it will often unconsciously move towards more complexity, causing delays and extra revenue. It's a bit like the quote by George Orwell who stated in his book 1984: "*The war is not meant to be won, it is meant to be continuous*". Similarly, for many IT suppliers the project is not to be finalized, but to go on forever.

Software systems are like diets
Thinking that other, modern technology will generate success is often a useless distraction. We tend to ignore our own failures but look instead at new promising technologies to solve our problems. Should we replace this database for that database? Should we have programmed it in Python rather than in Java?

The problem is that people tend to look for external solutions instead of taking a hard look at themselves. New technologies often promise the world and are therefore very alluring. But technologies are like diets, whenever a diet fails to result in significant weight loss we are inclined to think we need another diet, instead of simply adhering to the current diet.

"One of the biggest mistakes smart engineers make is optimizing a thing that shouldn't exist"

Elon Musk (1971)

The 4 project constraints

Every project has four traditional constraints. Time, Money, Scope and Quality. These limits are set by us at the start of the project and we are supposed to remain within them.

The Paperclip Mindset goals do have some overlap with these constraints, namely Time and Scope (called *Target*).

The main and important difference between the traditional four constraints and the Paperclip Objectives is that the first tells you to not go outside of the boundaries, whereas the latter incentivizes you to minimize the four constraints even more.

Okay, so let's see what the four Paperclip Objectives are.

"Perfection is achieved,
not when there is nothing
more to add, but when
there is nothing left to take
away"

Antoine de Saint-Exupéry (1900 - 1944)
Author of *The Little Prince*

The 4 Paperclip Objectives

Most people seem maximalists rather than minimalists. So instead of staying within the project constraints, Paperclip advises to approach it from the other side, to minimize the constraints. Just like the French author Antoine de Saint-Exupéry conveyed by his quote on the previous page.

"Work expands so as to fill the time available for its completion"

(Cyril) Parkinson's law
British management scholar

1 - Minimize time

To limit time is to become more decisive.

"Time Is the Ultimate Currency" Elon Musk

I like to quote Elon a lot. What he has achieved with SpaceX can only be achieved by fighting complexity, by reversing rocket science. Musk also said: "The best part is no part".

Parkinson's Law
Elon Musk often announces highly ambitious projects with seemingly impossible time frames. And that is a good thing. He probably is aware of Parkinson's Law, which states that work expands so as to fill the time available for its completion.

Minimizing time is an excellent way to foster decisiveness. It forces you to define a reasonable goal and it prevents being distracted by trivial stuff.

It also incentivizes people, for they feel the breath down their neck. Would you start working hard for a deadline two years in the future? If it's about a moon landing maybe, but not for most IT projects.

Analysis paralysis
Allocating fixed time periods (time boxing) reduces 'Analysis paralysis'. My own rule is "Functionality hard, Deadline harder", or in Agile parlance 'Fixed time, Flexible scope'. Which means that it is better not to deliver on all planned

functionality than to overshoot the deadline. For being allowed to overshoot the deadline causes the team to lose urgency.

"A goal without a deadline is just a dream." Robert Herjavec

Short deadlines
Short deadlines prevent the so-called Student Syndrome, which means 'doing the work until the last moment'. When we have a deadline for Sunday, we tend to wait till Saturday to really start working on it, at which point we realize it takes more time than expected. Therefore it is best to create short deadlines, as to create more pressure points and more reality checks.

Schedule a Launch Party
This advice can hardly be overstated. Just setting a deadline is not enough though, for it often creates *the illusion* of pressure. I have witnessed many projects in which deadlines were joyfully overrun, without consequences or shame. Internally the project team can always find 'credible' excuses for not meeting the deadline. But real pressure is created when a customer, an investor, or external individuals expect tangible results. A Launch Party that is widely announced in advance will help tremendously to create *real* pressure.

Pressure brings purpose and joy
The most enjoyable and successful projects I have realized were the ones that had to be done in a few weeks. The joy and the sense of accomplishment it brought to both myself and the other team members was invigorating. Meaningful progress seems to be one of the best motivators, even more than money and also more than receiving praise. People desire meaning

and purpose, and seeing results and reaching a target provides that.

Ray Croc, the man who grew McDonalds, said: "Happiness is the by-product of achievement". I often think of his words when meeting the deadline.

"There are no solutions.
There are only trade-offs."

Thomas Sowell (1930)
American economist and author

2 - Minimize target

Minimizing your target fosters focus.

If you want to achieve ambitious goals you can't allow non-essential parts to distract you. Most IT projects evolve towards a Christmas tree. More and more ornaments are gradually added to it in the name of quality, future efficiency or customer convenience. In IT this is often called *Feature Creep* or *Scope Creep,* the continuous or uncontrolled growth of a project's scope. The original scope easily moves out of sight, and so does the deadline. Therefore we need to constantly focus on minimizing the target and focus on the essentials.

Minimal Viable Product (MVP)
Identifying a Minimal Viable Product (MVP) is very helpful in preventing Scope Creep. An MVP is that with which you can 'get away with', negatively speaking. However, this is not enough and is no guarantee for a stable scope. You need to also define Proof of Concepts (PoC) and a prototype. Many projects have essential *Unknown Knowns* which need to be known as soon as possible by building a PoC. This helps in preventing sudden future delays.

A prototype is like a car without an engine and wiring. Functionality might be simulated. You can sit in it and get a feel of what the end result will be. In software terms this would mean a set of screens that are linked, with buttons and texts, and that a user can navigate through.

YAGNI (You Ain't Gonna Need It)
An important aspect of the Paperclip Mindset is YAGNI, which is an already existing acronym for You Ain't Gonna Need It. In hindsight you can notice that you didn't need it, but just knowing and remembering this term helps in realizing it in advance ("If you can't name it, you can't tame it").

Take for instance web hosting. I have seen projects in which the team thought they needed serious and expensive hosting from the start (including load balancing!), in expectation of huge loads. However, for most startup projects simple hosting is sufficient.

Creating luxury problems
Any project needs to focus on creating Luxury Problems. What is a Luxury Problem? For startup web projects it's like having too many customers signing up so that the IT system can hardly accommodate it. Often people try to solve scalability before any customer has signed up and this can take weeks, if not months, to implement. This effort could better be spent on releasing earlier and *attracting* customers instead of *supporting* potential customers.

Any startup tries to discover what functionality is marketable. Therefore it should better just focus on creating desirable functionality for a few users than to already prepare for possible market success.

Functionality centric
Every project is continuously distracted by the lure of additional functionality, from Essential to Nice-To-Have or the

outright Luxurious. Many of these suggested additions are cute or beautiful and therefore very tempting to include.

However, they are often not functional or don't add to the basic user experience. It seems like a no brainer, but one should keep focussing on essential functionality, not colors or animation. I often experienced that projects suffered major delays because unessential gadgets were implemented.

Focus on user experience
It is best to focus on building functionality for the frontend user, not for the backend user (reporting, maintenance). The backend functionality is secondary, and can often be done manually. The backend functionality only becomes important when enough users have signed on.

"Efficiency, which is doing things right, is irrelevant until you work on the right things"

Peter Drucker (1909-2005)
Management educator

3 - Minimize team

Minimizing the team simplifies communication and planning.

The best IT projects I ever did consisted of three or four dedicated team members, sitting within hearing distance from each other. It is not only efficient but also motivating, rewarding and pleasant.

We tend to think that when we know what, when, where and how to do it, nothing stands in the way of it being done. But that's not enough, for like author Derek Sivers once said: *"If information were the answer, we'd all be billionaires with perfect abs"*. One important element is missing, namely motivation, and working together in a small team is very motivating.

"Adding manpower to a late software project makes it later."
Brooks's law from Fred Brooks' *The Mythical Man-Month*

Working face to face
Working in the same room fosters responsibility, a shared purpose and mutual understanding. Having a small team means better communication and planning. In general: to meet someone is better than to call and to call someone is better than to text or email, although they all have their specific advantages.

Personally, I have completed tasks in a remote setting over the course of a whole month, while I knew they could have been done in a few hours when sitting side by side. Working remotely can be utterly frustrating and disheartening. Remote communication is often difficult, you can't just simply point to something on your screen. You neither pick up on many important signals, microexpressions and cues from your co-workers.

Projects require a thousand little decisions to be made, which are ideally tackled in minutes, without texting or calling, or having to email them in elaborate prose. By phone, people don't like to be interrupted often. However, we generally find it much more acceptable to be interrupted by a co-worker sitting next to us or while chatting at the coffee machine.

Imagine how frustrating it would be if that co-worker would have to schedule a Zoom session every time he wants to check if he is on the right track. Working in a small team builds social credit and a sense of comradeship, while working remotely builds misunderstanding, distrust and frustration.

In a small group the team members also feel more responsible for the group effort and can not so easily hide behind other members for failing to meet their targets.

Select multi-skilled individuals
If every task could only be performed by a separate dedicated individual it would require many people for all the different tasks that a project required. The communication between all the team members would quickly balloon. So in order to minimize the team it is essential that it consists of multi-skilled individuals.

Having fewer team members reduces the need for memos and meetings, which is a huge time saver. Elon Musk once said about his company SpaceX: "Everyone here is Chief Engineer" indicating the importance of individuals having a broader view.

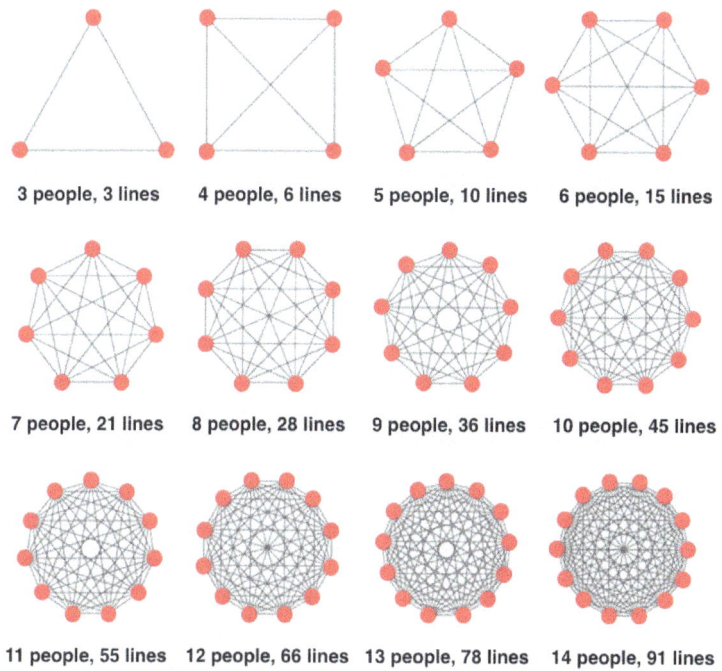

Exponential growth in complexity

Customer present

Ideally, the team works at the customer location. Working face to face with the customer is conducive to better understanding and easier communication. In these days of the internet it is tempting to think everyone can work remotely. Although it does have its advantages, the drawbacks generally outweigh the pros, as I have listed in the table in the chapter "Downside of working remotely". Customers generally also like their contractor being present, for it gives them a sense of control and an insight into what is really going on.

"The number of people having any connection with the project must be restricted in an almost vicious manner"

Kelly Johnson (1910 – 1990)
American aeronautical engineer

4 - Minimize technology

Minimizing technology means less bloat. But if you work with techies this is a difficult goal to attain.

For don't people just LOOOOOVE technology? We tend to think more technology is automatically better. But therein lies a great danger. For if you love technology, you tend not to tame it but to expand it. The complexity monster will then eat both you and the project for breakfast.

The curse of the current trend
IT has its trends too and I pretty much stopped following them. The problem with most new development tools is that the cost of switching to them is often higher than the benefit they would provide. If you are very familiar with database X the costs of switching to database Y are generally big. Database Y needs to really be much much better to justify the shift.

Sticking with the devil you know instead of falling for the new craze is generally advisable. It is probably wisest to use a few standard development tools that you know well. Remember that HTML, CSS, and JavaScript are web standards and programming languages like Python and PHP and a database management system like MySql are *de facto* standards used by millions. They will outlive any fashionable framework, library, preprocessor and over-hyped fad that comes along. Sticking to a few standard development tools also makes it easier to hire new team members since their resumes do not need to contain a rare mix of technologies.

Avoiding tech stacking

When needing some extra functionality it's tempting to install some plugin or library that is supposed to do just that (and often much more). However, this can be very dangerous for the project because you might enter the PlugIn-Hell. Since there are hundreds of plugins for anything you might like, it is easy to overload the system. Many Wordpress sites have been hacked because of faulty plugins with known security holes.

But plugins can cause other problems too. When upgrading to the new version they might clash with your other plugins, and cause the whole system to fail. When installing a plugin you basically add a black box to your system. It has input and output but its inner workings remain unknown. Whenever the plugin fails, you have to use trial-and-error techniques and reverse engineering to fix the problem, which is often very hard. In essence, you give up control when adding a plugin.

Once I was asked to help in releasing a delayed Magento e-commerce project. The thing with Magento, which is open source, is that it is such a complex monstrosity that adapting it is virtually impossible. You therefore need to rely on commercial plugins to provide even the most basic changes. Since many plugins were installed on this particular Magento system it caused enormous dependencies, flaws and limitations. The whole system had become one big black box. The team constantly needed to try and guess in order to fix ghostly problems.

Installing plugins is a form of tech stacking, which is best avoided. Stacking software tools eventually leads to loss of control which will lead to a failed project.

Favor quick manual implementation over automation
Often, something just needs to be done one time and in a flexible manner. Manual implementation then trumps automation. Automating the task can take weeks or even months but doing it manually can often be done in five minutes.

Favor non-technical solutions over automation
For instance, sometimes it is better to let users correct their mistake instead of letting the software prevent the mistake. Some 'bugs' or missing features are just not important enough to fix.

"One of my most productive days was throwing away 1000 lines of code"

Ken Thompson
American pioneer of computer science

Why time is the best constraint

Traditionally every project has four common constraints: time, scope, quality and budget. Which one would be best to minimize? Well, my most successful projects were constrained by time, and in hindsight I can clearly see why that is.

The great thing about the time constraint is that it constraints all others. Having limited time prevents one from spending too much money. One could of course argue that, because of limited time, one is compelled to hire a 1000 team members to speed things up, and therefore would still overspend. But this is generally not what happens, for time also limits spending. Time constraints also automatically limit the scope of the project.

Limiting the budget is great, but doesn't work as well as limiting time. Suppose you would limit the budget to 100.000 dollars. In that case the project might still take a year to finish it. The project would lose its momentum and team members would become frustrated, dispirited and could even change jobs. If you limit the 'quality' or 'scope' constraint, the same thing would happen.

The time constraint is the only one that keeps other constraints pretty much in check. And as an extra bonus the early results motivate all the individuals involved. Additionally it allows for user and customer feedback to be received early in the project. Valuable lessons can thus be learned.

If we have too much time we fall victim to the 'Law of Triviality', once coined by management guru Cyril Parkinson. The law argues that people within an organization tend to give disproportionate weight to trivial issues. Limiting time increases the pressure to deal with the important issues.

'Good Enough' is Perfect (for now)

'Quality' is one of the four standard project constraints. We tend to think we need high quality software and that the drive for higher quality or perfection is necessarily a good one. Anyone not disagreeing might be lazy or slacking.

But the drive for perfection is often not your friend but your enemy. Higher quality is not free but comes at the cost of time, budget and effort. Economists speak of "opportunity costs": when you decide to do X you can't do Y at the same time and with the same budget.

Perfection is the enemy of the deadline. I have experienced many projects in which the continuous desire for higher quality caused frequent release postponements. Since every release is a learning opportunity, postponing it is a delay in understanding the user or customer.

Perfection is unachievable anyway, it's like the horizon. When we move towards it, it moves away.

Optimal versus perfect
We have to learn that everything is a tradeoff, a compromise. When we would value quality as our highest goal we would require all cars to be a Mercedes or BMW, even when just going shopping. Although this sounds generous and ambitious it prevents many people from going from A to B in a Volkswagen which they would be able to afford.

The 'good enough' attitude is often an unpopular one, for it seems lazy. 'Good enough' is optimal, and optimal is actually perfect. But perfection is surely not optimal, a bit like the famous photographer Chase Jarvis once said "the best camera is the one that's with you".

The importance of small steps

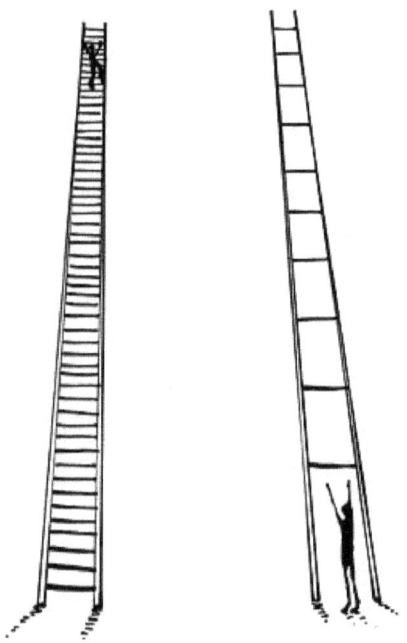

The road to perfection is filled with 'Good enough' steps. Every time we release a Good Enough milestone, product or service we allow ourselves to learn through user feedback and course-correct on time. But if we postpone a release for lack of perfection we deprive ourselves of those lessons and progress slower because we went off course. Since many IT projects nowadays are web-based we can iterate often. We don't need to release CDs but can try new functionality, correct mistakes and fix bugs by simply uploading the new code.

The benefits of working in-person

IT projects seem ideally suited for working remotely. Since no physical stuff is created, all the people involved need not be physically present. But there are enormous drawbacks to working remotely. In order to get things done quickly it really pays to work in the same location.

Having an IT team that is geographically dispersed does have certain benefits. One can select professionals from a world wide pool of talent. Individuals can work from home and save on travel time and expenses. But it also brings many disadvantages that are often too abstract to notice or measure.

Savings in travel costs can be easily measured, but increases in motivation, emotional connection and mutual understanding are hard to observe, to compare, or even to quantify. For years I have observed and compared these parameters and noticed that the differences are huge.

My conclusion: if you want to achieve a lot in a very short time, collaborating in-person is almost the only way.

In the table below I have indicated my perceived advantages and disadvantages of working together in-person (physically present) or remotely (videocall, email, phonecall, chat, etc).

	In-person	Remote
Mutual understanding	*****	*
Building Social Credit	*****	*
Control	*****	*
Trust	*****	*
Need for Administration	*****	*
Motivational	*****	*
Building a Shared purpose	*****	*
Turnaround time	*****	*
Travel savings	*	*****
Human resource pool	*	*****
24/7 support	*	*****

More stars indicate more benefits.

Working in-person fosters trust for one to see that the other person is really working on the project and not spending time on his hobbies while working from home. It also creates more control, for it is easier to notice if someone is working on an issue for too long and needs assistance.

Elon Musk seems to have the same appreciation for working in-person. After the Covid pandemic, during which many people worked from home, he sent out an email to all Tesla employees which stated:

Everyone at Tesla is required to spend a minimum of 40 hours in the office per week. Moreover, the office must be where your actual colleagues are located, not some remote pseudo office. If you don't show up, we will assume you have resigned.

The more senior you are, the more visible must be your presence. That is why I lived in the factory so much - so that those on the line could see me working alongside them. If I had not done that, Tesla would long ago have gone bankrupt.

There are of course companies that don't require this, but when was the last time they shipped a great new product? It's been a while.

Tesla has and will create and actually manufacture the most exciting and meaningful products of any company on Earth. This will not happen by phoning it in.

Thanks, Elon

Billionaire entrepreneur Peter Thiel, with whom Musk worked at Paypal, understands the benefits of working on location too. Allegedly, he offered staff an extra $1,000 a month if they lived close to the office.

A simple simplification technique

Whenever we run into serious problems and delays while realizing a task or project it is generally counter intuitive to look back and identify if you took a wrong turn somewhere. We like to look forward when solving a problem. But taking a step back can be tremendously helpful.

Whenever a problem turns out much harder to solve it's good to ask yourself and others three simple questions:

1) Do we REALLY need it?

Is it part of the essentials or is it nice-to-have? Often, it turns out to be the latter or even pure luxury. Luckily an acronym already exists for this question, namely YAGNI, which stands for You Ain't Gonna Need It. During a project all kinds of nonsense is gradually added to the project and this way you can get rid of it.

2) Do we REALLY need it NOW?

Ok, so if we REALLY need it, do we REALLY need it NOW? This is also a very powerful question to which the answer is remarkably often "No". By postponing it you can save yourself lots of time and focus on more essential things.

3) Do we REALLY need it NOW and LIKE THIS?

Ok, so we REALLY need it NOW. But do we also need it LIKE THIS? Can it be done manually or does it need to be automated? Does it need to be so fancy? Can we think of a shortcut or a workaround? It always turns out that there are so many ways to skin a cat. The simplest solution is often hardest to find, but contemplating its various options for half an hour can result in saving days of work.

The best way to ask these three questions is to do it groupwise and challenge each other to say 'No' to any of these questions, but still remain realistic.

The importance of pushback

Napoleon Bonaparte once advised: *"Never interrupt your enemy when he is making a mistake."* Conversely, I would advise: *"Always interrupt your customer when he is making a mistake."*

Customers often come up with new requests for functionality. We tend to readily accept the validity of their request because we think '*Your request is my demand!*'. We consider the customer to be king, but we also understand that extra requests will nicely result in higher revenue. However, this attitude is not conducive to project success.

Customers often request *What* they want (the functionality) without explaining the *Why* (the problem). Therefore it is useful to push back and ask for further clarification.

1) "What is the exact problem that you are trying to solve?"

For the exact same problem you can often come up with a different solution. I once had a customer who requested help with implementing some functionality. We discussed it for ten minutes, after which I really understood what and why he wanted it. I then realized he had unintentionally led me to the wrong solution to his problem, so I suggested a far simpler solution. This made him very happy because it created huge time savings.

A problem well stated is a problem half solved. Einstein is

reported to have said that if he only had one hour to solve a problem he would spend 55 minutes defining the problem and the remaining 5 minutes solving it routinely.

2) "How often would this solution be used?"

Ok, so after question one is satisfactorily answered you can ask another one. How much time and effort will this solution save? You can both agree on the problem and its solution, but that doesn't mean it's a solution worth investing in. Maybe it solves something that only a few people need or is only required a few times a month. Maybe it can be more easily done manually instead by either the users or by the technical team. I have often experienced that doing it manually for the time being is a satisfactory temporary solution until the frequency rises to serious levels.

You might think that customers don't like your pushback, but my experience is that they actually do. Of course I tell them also about the many challenges we are facing and that we need to keep the essentials in focus. I also introduce them to concepts like Scope Creep, Scope Freeze, and the Law of Triviality. When customers understand the dynamics of the project, they tend to be far more sympathetic to you pushing back. They come to understand that you are trying to better solve their problem instead of slavishly developing the functionality. Which results in faster and cheaper solutions.

"Football is a very simple game, but the hardest thing is simple football"

Johan Cruyff (1947 – 2016)

Cases

Everybody agrees with Einstein's statement "Everything should be made as simple as possible, but not simpler". However, it's very difficult to know what the simplest implementation exactly is.

We often suspect that our system is too complex, but we don't know what to leave out. Unfortunately there is no simple formula for finding the simplest solution, but luckily you can learn from examples. I have therefore described some cases below. Some are a bit technical but others can also be understood by laypersons.

Case 1: Live now, pay later

For a startup web service the customer considered it essential to select a payment provider in order to allow visitors to pay for their service.

Integrating a payment provider to your website is generally not very difficult to do, but it still takes time, effort and money. However, time was running very low and we needed to put stuff on the hacking block.

So I suggested skipping the whole payment page. Considering the low volume of payments (and high price) it might be best to leave it out. We would simply build a form with which they could order the service.

After an order would come in they would then call and email the customer and provide bank details for the actual payment.

In a low volume situation, which many startups are in, this gives the added benefit of getting to know the customer. With the excuse of calling we can ask him things that we normally would not be able to know.

After some discussion we wisely chose the 'doing-it-manually for the moment' solution, which saved money, time and effort. If sales volume would really turn out to grow a lot (a luxury problem), it would justify the investment to integrate a payment provider interface.

Case 2: Stop automating everything!

How to simply save a man-month.

As part of a much bigger web project, a retailer wanted to create a Google Map with the locations of its two hundred branches. The locations were stored in a Microsoft Dynamics system, separated from the web server.

The technical team decided to build a daily data feed from the Dynamics server to the web server. This is of course an ideal solution, but ideal solutions are not optimal solutions.

This is a typical example of doing something by automation rather than by hand. It took about a month to implement.

As the saying goes, "If the only tool you have is a hammer, you tend to see every problem as a nail". Similarly, when you are an IT professional, you tend to think everything needs to be automated.

But there is a simpler alternative. As you know, stores don't change location very often, they don't have legs. And unless you're Starbucks or Burger King, new stores are not added on a daily basis.

It would have been much better to just download the data once and create the map. Could have been done in a day's work.

By choosing to build an automated data feed the company spent 30.000 euro and lost valuable time. And mind you, that was at the beginning of the project, so they wasted valuable time on something non-essential.

Case 3: Fake it till you make it

How IBM saved millions by NOT building a PROTOtype.

This is not an example from my personal experience, but certainly worth mentioning, since it explains very well how far back one can go in testing a new product on the general public.

Marketing research by IBM showed that many people would be very interested in using speech-to-text software, with which a user could simply talk to a computer instead of typing text.

Our initial impulse to test the product would be to first develop the software for it, after which a test panel could try it out. But wait, isn't there another, easier way?

Yes, there is. IBM wisely decided to NOT build the software, which would have taken a few years at least. Instead they chose to test out speech-to-text with regular test candidates and a real typist, hidden from view. This is called Pretotyping (from 'Pretending'). Beside the obvious time and cost savings, the great advantage was also that the typist was probably much better at understanding speech than any software at the time could have been.

Surprisingly, it turned out that people didn't really like speech-to-text software, even though many had indicated enthusiasm beforehand. This is understandable, a speech interface for a regular text document might be reasonably convenient, but controlling an application is another thing. For instance, telling the word processor to go two words back and correct a particular text can be cumbersome. IBM quickly learned that marketing research doesn't always render reliable results and that developing software that users can really use is often extremely difficult.

Case 4: Who needs Excel?

How 'thinking outside the box' can save money.

For a start-up company I worked for we received various Excel files with around 40.000 items. They needed to be imported into the database.

Initially, many would decide to use the Excel function library of their particular programming language. So did I. But since Excel sheets can be very complex it turned out that just reading an Excel file would create a steep learning curve. And indeed, the programming library contained many different functions for so many different situations.

The workaround was very simple. Since the data in Excel had one row for each item, converting each of the 12 files to the CSV format (Comma Separated Values) made it way easier to import. No library needed at all! Time saved: probably two weeks.

You might now think "Okay, that's easy. I would have come up with that solution.". You might be right, but in my experience people often do not even consider a short-cut or a workaround when confronted with a problem like this.

Case 5: Forget the MailChimp API

Another case of a manual solution instead of automation.

A young start-up wanted to send a newsletter to subscribers via MailChimp. For that, it was suggested we needed an API connection between our own user database and MailChimp. MailChimp provides an API through which you can synchronize the data. It was somehow suggested that we could only send our first newsletter once the API link was implemented.

But was that really necessary? Certainly, an API link would be ideal. New users in our own system would automatically end up in the MailChimp subscriber list. And vice versa, those who would have unsubscribed could be marked in our own system.

But building the API would probably have taken at least a few days. Luckily, there's a much simpler solution. At that stage, the newsletter only had about one hundred subscribers. And the newsletter had a low frequency of once every month. Doing it manually was the simplest solution and it could be implemented within an hour. Just export the data from the database and upload it to MailChimp. Those who unsubscribed needed to be manually marked in our own system. Done!

The general lesson is here: don't build solutions for non-existing problems. Create luxury problems instead, which means: attract so many subscribers that an API is desirable. The API would be great when having ten thousand subscribers, but not just one hundred.

Case 6: Beware of 'Layer Love'

How striving for technical perfection caused me to swim through peanut butter.

Years ago I was recruited by a Dutch hosting company to work with their new framework called *Mody*. *Mody* was their backend system which mixed database data with HTML and sent the result to the browser. The new system was to replace its current but outdated system, held together by tape and glue.

For me the *Mody* project was both difficult and amazing. 'Difficult' because it was so frustrating to work on, and 'amazing' because it taught me many valuable lessons, especially on how *not* to design a system.

Careless about costs

My first lesson was that most technical staff do not seem to care about costs. *Mody* had already been in its 'final stage' for 2 years (!) when I started working on it. And it hadn't been released yet, which rightly got management worried.

The techies working on *Mody* didn't mind the delays however, for the company was making good profits, and they were working on this wonderful and very promising system that would certainly justify all expenditures. However, I got the impression that they were in fact exercising their personal hobby, their pet project, at the company's expense. This total disregard of technical staff for financial costs is something I have witnessed more frequently.

The love for Lasagna Design

Another lesson was the love of layers many software engineers have, like the layers in lasagna. Getting introduced to *Mody* was like a culture shock to me. The designers seemed to have gone to great lengths to make it as complex as possible. And they succeeded! In a minimal system the backend just consists of two layers: the database and the software (in their case PHP). The software interfaces with the database and mixes it with text and HTML. However, the techies had gotten on a layer creation spree. *Mody* consisted of no less than 9 layers, all adding to obscurity and complexity. It incorporated several new formats and techniques, with impressive names like SOAP, XML and XSLT (all for coughing up just a rather straightforward HTML page).

Working with it was both time consuming and a debugging hell. This book is particularly meant for non-technical people, so I will spare you most of the technical details. Because of the many layers one could not easily determine in which layer something went wrong, for each layer generally didn't return the errors to the subsequent layer. It was like the technical version of the Chinese Whisper. One can also compare it to having nine different company departments passing on a simple customer order before it reaches the warehouse, and all the way back again for shipping the product to the customer address. What could possibly go wrong!??

I think, as a rule of thumb, that whenever a system consists of more than 4 layers, it is time to pause and reflect and try to lower it.

Techies don't push back
One other insight that this project taught me is that techies generally don't push back. Management had issued two requirements at the start of the project: *Mody* should have multilingual capability and the database software should be agnostic. Instead of advising management to reconsider these two requirements, the builders went all out and created a complex system to accommodate them. This must have delayed the project quite a lot. Years later, I found out that both features were never used, they used just one language and stuck with the original MySql database. It was a typical example of YAGNI, *You Ain't Gonna Need It*.

Whenever a customer says they want some functionality you should almost always push back. Do they really need it, isn't there a simpler way? Generally it turns out that the customer had just let his heart speak instead of his brain.

Beware of *The theory of everything*
Once you give techies lots of time to create a new framework, they will end up trying to create the software version of *The Theory of Everything*. I have experienced it myself too when designing a new Content Management System: one quickly falls into the trap of feeling it should be everything to everyone, it should accommodate any future functional demand, no matter how unlikely. And before you know it you don't think of the deadline anymore, but of creating this groundbreaking and unique system that might change the world and will make everyone admire you for designing it.

With *Mody* this happened too. The final system was supposed to be fantastic and all encompassing. But it wasn't. It was slow and complicated and it took 5 to 10 times more to get simple functionality done. It was truly like swimming through peanut butter.

So this tendency to create the *Wonderful System for Everything and Everyone and All Times* needs to be acknowledged early on and be tamed.

The power of the sunk costs
Three weeks into my involvement I warned management on the risks and dangers of *Mody*. Initially they were shocked but wisely decided to compare my views with those of the designers. Unsurprisingly, they decided to safely side with their IT staff. Since management had no technical background they rather trusted those with the good news. I can't blame

them. However, they must have felt there was probably something wrong with the system already for it was way over budget and time.

I had warned that technical staff would resign from the company, just for not wanting to work with this behemoth of a system. And indeed, a few years later I was told that this had actually happened. And a few years after that I heard the whole system was scrapped and replaced by a new one (probably just as complex).

In closing, this case is not so much about a particular goal that could have been achieved much simpler, but an example of making all the wrong design choices and of hardly enforcing any deadline. If you want to build a rocket to go to Mars, two years is a short period. But if you just want a new framework, I would say to release the first version within 2 to 3 months and improve gradually.

Case 7: Who needs an API?

How to save a year of work by choosing a simple solution.

A company wanted to grant a customer *limited* access to its main MySql database. The customer could then autonomously query its own data, which would save time and money.

The technical team decided to build a proprietary API (Application Programming Interface). The customer could then write specific software to query the data through use of the API.

Building the API must have taken at least six man-months. Eventually, the total code size grew to 500Kb, which is huge in my view. After a year, the customer had still not made use of it (and maybe never has).

Learning to interface with such a proprietary API takes quite some time. This new knowledge can not be used elsewhere either, so it is by default not a very profitable investment.

Luckily, there's a simple alternative though. Many IT professionals are already familiar with SQL, a generic database language that MySql also listens to.

Therefore, the company could have saved quite some money and time by just offering a so-called VIEW on the database, which limits the amount of accessible data. This way the customer could easily query the database via SQL and various MySql tools like phpMyAdmin. Time saved: half a year!

Case 8: Cutting out the nonsense

On how to save time by building a Proof of concept.

Once I worked for a company that replaced its existing webshop Magento 1.0 with Magento 2.0. Everybody who has worked once with Magento knows it is huge in size (500 Mb) and can be very slow.

The slowest part of webshops is often their 'select and search' product page. Visitors click on check-boxes to narrow down the search results. This particular Magento webshop contained only about 3000 product variations. On a good day, it took at least 5 seconds for the search results to appear, even with one visitor on the site! This was acknowledged as a serious problem, since slow page speeds can easily cause customers to leave the webshop out of frustration.

Out of curiosity, I just wanted to know how fast this search process theoretically could be, once you simply aggregate the product data. So I built a Proof of Concept. By doing so, to my surprise, I noticed I was inclined to overcomplicate it in various ways.

1) Initially I thought, stupid me, I first needed to build a database to store the info, just like with Magento. I would then create a process to aggregate the data. But this was silly because I needed no database at all, I could just use a multi-dimensional, hard-coded product array to simulate it.

2) There was no need to aggregate the data either, for I could start with aggregated data to begin with.

3) I also felt I needed real data to simulate reality. Stupid me again. Coming to my senses, I decided to mainly use gibberish data. I did however use real product names and real categories, for it feels better and wasn't a lot of work.

4) I also thought initially I needed to show real product images, but quickly realized that was not necessary at all. I could just show the coded data on the search results page.

A Proof of Concept generally needs no real data, no database, no fancy user-interface and no internet connection.

On the basis of the product array, I wrote two programs:

1) A small program to generate 22000 random product variations. Each product variation had ten semi-random characteristics like brand, price, size, category and weight. These random products were stored in a simple CSV file.

2) Then I wrote a program to create a user interface, scan the CSV file and select and show the qualifying product variations.

It all took about two days of work to write and fine-tune the small programs. The result can be tested on https://paperclip.one/productfilter/. It turned out that even on shared hosting, costing only $10 a month, searching through 22000 product variations took less than 0.03 seconds!!! What an impressive test result indeed.

However, this PoC was never intended to be integrated into Magento. The company had to rent many servers at Amazon Web Service to speed up the site and to accommodate for traffic peaks, both during special sales and right after sending newsletters. Total hosting costs were probably in the range of $10.000 per month!

Case 9: Avoiding the time zone hell

How a geographically distributed team experienced misunderstandings and cost overruns.

Nowadays companies can inexpensively hire IT professionals in developing countries like India or the Philippines, often at a third of the hourly rate of local professionals.

It seems like an easy decision, the geographic distance obviously hinders communication, but allegedly that can be mitigated via video conferencing tools like Zoom.

However, my experience with working remotely has been remarkably negative, it does not seem to foster the trust, motivation and shared purpose that working face-to-face does. But admittedly, it's sometimes the only option available.

One of my experiences was during a project with an Italian development team, while the customer and I were located in the Netherlands. Even though the complete team held daily video conferencing sessions, getting a grip on the project proved far more difficult than when the complete team would have been physically present in the same location.

I clearly noticed the huge difference during the final release of the project. For that, the main developer flew over from Italy to Amsterdam for a few days. As always, many unforeseen problems arose during the release. But by quickly discussing the various solutions, these issues could swiftly be tackled. I can not imagine we would have successfully released if the Italian developer would have stayed in Italy.

"I am a big believer that people are more productive when they are in person." Elon Musk

Water cooler moments
I have experienced that working together in the same building, and preferably the same room, fosters a team spirit and enthusiasm that is difficult to create via Zoom.

Discussing problems and explaining ideas can be done more quickly and with less ambiguity. Also, interpersonal annoyances can be noticed early and subsequently resolved. During water cooler moments, ideas, opinions and jokes are shared that would normally not be brought up during video conferencing.

Since my various remote working experiences, I strongly favor working on location. It works wonders, since it drastically simplifies the project.

Case 10: Let's create luxury problems!

"Suppose we get a million users!" Or why choosing Amazon Web Services can be a foolish choice.

Let's create luxury problems! If you simply want a sandwich, you don't start off by building a kitchen first, just because you might want a thousand sandwiches later on! It seems a no-brainer, but many projects are approached in this way.

When people start a new project they often plan for perfection. They prepare for all kinds of eventualities and therefore start with strong fundamental foundations. Sounds professional, but it often distracts from the real issues.

Generally, the most important issues of any new IT project are not about scalability or security. But rather, do the essential parts work? And, did we deliver something the end user and the customer actually wanted and envisioned?

Creating success is more important than building solutions for possible future success.

Once market success is achieved, which is already quite a feat in itself, other issues like scalability, design and security should be addressed.

So, when starting a project, you don't need a full-fledged (AWS) configuration with load balancers, database replicators, a cluster of application servers, etcetera. All because of an unlikely scenario: "Suppose we get a million users!" You can, in principle, start with shared hosting. Or VPS. I have done so and even postponed registering a new domain name, I just set up a subdomain. It means that you can start the same day with releasing the first version. Customers generally like this practical approach.

I know this all sounds very unprofessional, but professionalism is often an excuse for very elaborate preparation (read: missed deadlines).

Case 11: Avoiding the flight forward

A start-up company gathered and analyzed social media data, which they stored in a MySql database.

The database grew steadily. It turned out that reports were kind of sluggish because the queries (the SELECT statements) executed relatively slow.

In order to solve this problem, the technical team decided to replace the MySql database with ElasticSearch, which is known for its fast queries. This shift must have taken quite a few man-months to implement.

This is a typical example of the flight forward (when experiencing resistance, move forward).

By replacing the well-known MySql database by the less known and very different ElasticSearch, it opened a whole new can of worms.

The ElasticSearch interface is completely different from MySql, so it requires new tools, new functionality and new learning. More importantly, it turned out that ElasticSearch had slower(!) inserts than MySql, which posed a new problem! Goodbye problem A, Welcome problems B, C and D. What irony!

Regarding continuity, ElasticSearch poses another problem too, namely that it's harder to find technical staff with knowledge of ElasticSearch than of a regular RDBMS like MySql.

Alternatively, the company could have solved the 'slow query problem' via simpler means. One obvious solution would be to buy faster hardware and more internal memory. This is generally very inexpensive compared to man hours.

Additionally they could fine-tune the indexes and tables. If this still would not have worked they could have aggregated the data for some slow queries. By choosing these solutions, they probably would have easily saved a few man-months.

"Money is not the 'scarce commodity'. Small, dedicated and talented teams are"

Elon Musk (1971)

FAQ

In what way is Paperclip different from Scrum and Agile?

Both Lean and Scrum describe a development process. They both have valuable aspects of which Paperclip shares a few, like simplicity, face to face development and MVP.

Agile describes a set of values that Paperclip overlaps with, like simplicity and working face-to-face. But Agile does not describe simplicity.

Paperclip is about implementing radical simplification. It's about removing unnecessary, untimely, and overcomplicated elements, both in planning, functionality, team size and communication.

Isn't Paperclip like the Lean Startup ?

No, although it shares some ideas, like being frugal. But Lean is about an iterative process, not a Mindset.

Lean is about Startup projects, during which it is often very unknown who the users will be and what they actually want. Unlike the Paperclip Mindset, Lean focuses on tuning the engine.

Besides project time, the Paperclip Mindset also focuses on team, target, and technology. The guidelines are even applicable to projects outside of IT.

The Paperclip Mindset and Lean can be applied in combination.

Isn't Paperclip another word for KISS?

No. KISS, or "Keep It Simple, Stupid!" indeed praises simplicity (in general), but Paperclip provides insights through examples and guidelines on how to achieve simplicity (specifically for IT projects).

Everybody agrees with keeping it simple, it's a no-brainer. But what simplicity actually is, few can explain. In my experience, many proponents of the simplest solution can still come up with extremely complex implementations.

Did you ever have a failed project?

Is the pope a catholic? I have been responsible for projects that experienced cost and time overruns. But this is almost unavoidable (think Hofstadter's Law). The trick is to keep these overruns within reasonable limits. Generally I succeeded because of my pessimistic attitude towards IT projects, but it is still really hard.

Regarding really failed projects: in 2006 I was part of a large project during which I felt I had to speak up and blow the whistle, warning of time overruns. To my frustration and amazement few cared to listen. So I left the project.

A year later, my pessimistic predictions even turned out to be too optimistic. The new site was launched one morning (half a year later than scheduled), it crashed within two hours and the old site was reinstated. It took another eight months (!) before the new site was successfully released.

What kind of project management software do you prefer?

People tend to think that the best Project Management Software will make all the difference. But I disagree. It's a bit similar to thinking that the best tennis shoes will make all the difference for your game.

I'm a proponent of simply using a spreadsheet (Google Suite). Everybody knows what spreadsheets are, so there's little need for instructions or for everybody to install specific software. Spreadsheets are also highly flexible. But more importantly, they force you to keep it simple and structured.

I learned that the elaborate film crew of the hugely successful TV drama Downton Abbey simply used colored Post-it notes on their planning board, like with a Kanban Board. This has a great advantage, namely that it's very visible for all people involved. From my own experience this is a great way to create motivation and a shared project awareness among team members.

What is your experience with employing the Paperclip Mindset?

It works very well, but is far from a guarantee for success. The remarkable thing is that I still notice how easily I can complicate things. But the good thing is that I now often recognize this behavior beforehand, so that I can limit the damage done. Writing down the various cases really helped me with this.

Besides the Paperclip Mindset I have also learned how to best approach a functional or technical problem. There are generally many ways to skin a cat, but the first thought-of solution is often the most costly and elaborate one.

So, what I do to get the simplest solution is to first stipulate all possible solutions and then discuss them with the product owner or a technical person, depending on the kind of problem. This can involve an algorithm, a user-interface issue or a database setup. I have noticed that by explaining the problem and discussing the potential solutions it was both easy and very satisfying to identify the simplest one.

Why so much emphasis on simplicity, why not perfection or technical excellence?

Perfection is the enemy of the deadline. Life is full of compromises and any IT project is no different. Think of the Pareto Principle (the 80/20 rule) in this regard. Getting to 80% of your goal is often relatively easy. Adding another 20% can easily be 10 times as costly.

Perfection should be about finding an optimal balance between wishes, costs, time and effort. That would be ideal for a customer. The Paperclip Mindset intends to help with striking that balance.

www.ingramcontent.com/pod-product-compliance
Lightning Source LLC
Chambersburg PA
CBHW070111230526
45472CB00004B/1216